Janaísa Luiza Cristino Lucas
Alex Aparecido Ferreira

Characterization and production of an electrical device based on graphene

Janaísa Luiza Cristino Lucas
Alex Aparecido Ferreira

Characterization and production of an electrical device based on graphene

The fundamental process for obtaining and applying graphene

ScienciaScripts

Imprint
Any brand names and product names mentioned in this book are subject to trademark, brand or patent protection and are trademarks or registered trademarks of their respective holders. The use of brand names, product names, common names, trade names, product descriptions etc. even without a particular marking in this work is in no way to be construed to mean that such names may be regarded as unrestricted in respect of trademark and brand protection legislation and could thus be used by anyone.

Cover image: www.ingimage.com

This book is a translation from the original published under ISBN 978-620-6-76226-3.

Publisher:
Sciencia Scripts
is a trademark of
Dodo Books Indian Ocean Ltd. and OmniScriptum S.R.L publishing group

120 High Road, East Finchley, London, N2 9ED, United Kingdom
Str. Armeneasca 28/1, office 1, Chisinau MD-2012, Republic of Moldova, Europe
Printed at: see last page
ISBN: 978-620-8-26312-6

TABLE OF CONTENTS

Summary

The general objective of the work we are carrying out consists of the process of fabricating devices in two-dimensional materials, including lithography, metal deposition for electrical contact patterns, morphological characterisation and electrical measurements. The sample was produced using a method called Mechanical Cleaving. This method basically consists of manually separating the stacked layers of a graphite crystal repeatedly by sticking and peeling the crystal to a piece of tape. This is then attached to the silicon dioxide substrate and pressed manually so that the flakes adhere to the substrate. The number of layers in the sample was determined by optical microscopy. The sample was also analysed using Raman spectroscopy. To manufacture the contacts, a *LaserWriter* LW405 was used for optical lithography, followed by Cr/Au deposition and a *lift-off* process to finalise the device. Our work aims to understand the basic characteristics of electrical transport, the device fabrication process, electrical measurements and morphological characterisation, as well as contributing to the first steps in research into the fabrication and characterisation of two-dimensional devices at the Physics Department of the Federal University of Viçosa.

Objectives:

1.1 General Objective:

The process of manufacturing devices in two-dimensional materials, including lithography, metal deposition, morphological characterisation and electrical measurements.

1.2 Specific objective:

Manufacture of electronic devices based on graphene, following these steps:

- Exfoliation of the material on SiO_2;
- Detection of mono, bi or trilayers;
- Electronic lithography (or light beam lithography, via *laserwriter* or photo-aligner, for example, in the case of larger lithographic patterns);
- Metal deposition for *source-drain* contacts;
- "*Lift-off*" of the samples to remove resistors and spurious metals/oxides.

2 Introduction

2.1 Carbon

Carbon is an essential element that makes up most of the raw material of life forms and is the basis of organic chemistry. It has four electrons in its valence shell and when it forms chemical bonds, it can have three forms of hybridisation: sp, sp^2 and sp^3 . We will be interested in studying the structure characterised by sp^2 hybridisation that characterises the material studied, graphene, in which there are three valence electrons that share σ covalent bonds in the plane, and the remaining electron, which occupies the π orbital, is weakly bound and is responsible for most of its electronic properties [2].

2.2 Graphene

P.R. Wallace theoretically studied graphene in 1947 [30]. But it wasn't until 2004 that Andre Geim, Konstantin Novoselov and collaborators from the University of Manchester (England) and the Institute of Microelectronics Technology (Russia) isolated a graphene monolayer for the first time, describing its manufacture, identification and electrical characterisation. They used a simple and very effective method: exfoliation. They extracted thin layers of graphite from a graphite crystal using adhesive tape and then transferred them to a silicon substrate. [21].

The first group to propose this method was led by R. Ruoff, but they were unable to detect any monolayers [31]. The group led by Manchester, with the help of AFM, managed to detect some fragments made up of just a few layers. Among these detected fragments, flakes were observed that had only a single layer, and so graphene was identified [21].

Graphene is characterised by a two-dimensional hexagonal crystalline structure of carbon atoms, in which it presents a honeycomb-like structure. This material has a zero *gap*, high charge carrier mobility and linear dispersion of the density of electronic states around the K and K' points of the first Brilliouin zone. There are several ways to obtain the electronic dispersion of graphene, such as *tight-binding* calculations [4] and first principles calculations [5]. Either method will give very similar results for the electron dispersion of the valence band.

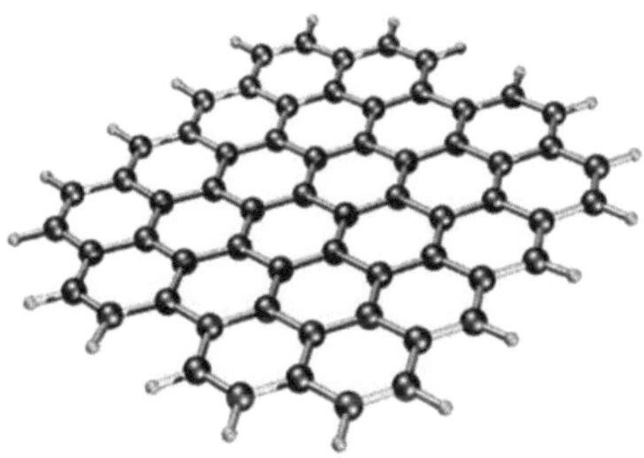

Figure 1Hexagonal network of a graphene monolayer, showing the carbon atoms organised in a honeycomb-like network [23].

The valence and conduction bands touch at only one K point. The energy gap between the conduction and valence bands is large at other points in the Brollioujn Zone. Therefore, the region of greatest interest for studying physical phenomena is the region around the K point. Analysing these regions closely, we see that they are similar to a cone. The conical dispersion approximation is good even for the energy range of interest for optical phenomena in the visible region and electrical transport ($|E| < 1eV$) [2].

Near the Fermi level, where there is no curvature in the electronic dispersion, we can obtain very interesting physical consequences. For example, when analysing the vicinity of the K point, the electrons obey the Dirac Hamiltonian for zero-mass fermions [6], i.e. the electrons are treated as if there were no mass.

MCG is restricted in its application in the electronic devices market because it has no energy *gap.* The opening of the *gap* is essential, for example, in the manufacture of field-effect transistors [2]. To overcome this difficulty, we can obtain the formation of a graphene bilayer (BCG), which is obtained by stacking two MCGs. This type of structure is a zero-gap semiconductor with high carrier mobilities. Unlike the MCG, the dispersion of the density of electronic states is quadratic in the BCG. When an electric field is applied perpendicular to the plane containing a BCG, it becomes a material with a non-zero *gap.* The width of the *gap* can be controlled by adjusting the external field being applied, i.e. the *gate* voltages applied to the fabricated device, which are related to the density of carriers in the graphene. The width of the *gap* is still small compared to other semiconductors (around 150 meV) and the application of BCGs in technological devices are ideas for the future.

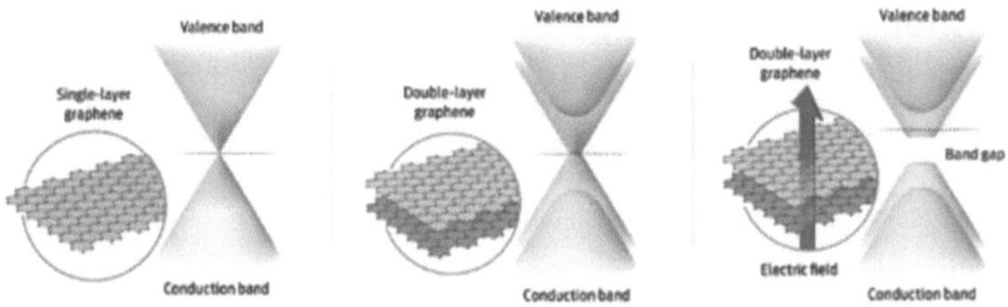

Figure 2 (a) Hexagonal network and electronic dispersion of a graphene monolayer. (b) Hexagonal network and electronic dispersion of a graphene bilayer. Note that there is no electronic gap. (c) Graphene bilayer under the effect of a perpendicular electric field. The width of the open electronic gap depends on the modulus of the applied field [24].

Another possibility for opening the *gap* is to oxidise [7, 8, 9] or hydrogenate the graphene [10, 11], which would cause the carbon atoms to hybridise. However, when electronic devices are made from graphene oxide or graphane, they show low electron mobility and some of the properties seen in pure graphene are also not observed [2].

The quantum Hall effect is another electrical measurement that characterises a graphene device. By applying an intense, perpendicular magnetic field to a graphene bilayer where the charge carriers move, the energy dispersion of a two-dimensional system unfolds into discrete levels. These discrete levels are known as Landau levels. The quantum Hall effect in a BCG sample is different from the effect for an MCG and for a traditional semiconductor. Conductance plateaus in the bilayer occur for integer values of the conductance *quantum*. However, there is no zero plateau as there is in a traditional semiconductor. It is therefore possible to distinguish a graphene monolayer sample from a bilayer sample by the quantum Hall effect [3].

2.3 Other materials with a similar structure to graphene

In addition to graphene, there are other materials that have a similar structure. For example, hexagonal boron nitride (*h-BN*) is a material that has a hexagonal structure, as mentioned for graphene. However, *h-BN* is an insulator, with a gap width of approximately 6.0 eV. As this material has a smoother surface than SiO_2 (after exfoliation), it has less charge build-up, *dangling-bonds*, etc. It could also be the basis for developing ultra-thin devices. If the graphene is deposited on a layer of h-BN, there is less possibility of scattering, and its electronic mobility will be higher.

Exfoliation of transition metal chalcogenides is also possible. Of particular note is mobilidene disulphide (MoS_2). This two-dimensional material has a well-defined direct gap of 1.9 eV. Its use would be in the manufacture of electronic and optoelectronic devices on ultrafine scales. Various studies are being carried out on the application of this material. There are numerous other materials, such as $MoTe_2$, $MoSe_2$, WS_2 , etc, but our focus will be on graphene.

Using the transconductance curve (ISD x VG), we can obtain another important electronic property: the electronic field-effect mobility (μ) of the charge carriers. This property gives us an idea of the quality of the devices manufactured, because the greater the number of impurities in the device, the lower its electronic mobility [12].

2.4 Properties of graphene samples

The transport-related properties of graphene samples have been studied recently. Graphene has been isolated on a SiO_2 /Si surface [13], due to its clarity when observed via optical microscopy [14]. The quality of the graphene samples supported by the substrate is related to the quality of the substrate. Therefore, the choice of substrate into which the samples will be inserted influences the quality of the samples and, consequently, this is related to the electronic properties of the graphene [1].

The graphene samples are adapted according to the substrate used. When a SiO_2 substrate is used, which has a roughness of approximately 3Å, the graphene sheet adapts to this and a very smooth sample is not presented. As a result, the graphene will show irregularities in some places, making electrical transport difficult. Another aspect that influences transport is the impurities present in the substrate. The contamination of the substrate by external atoms during preparation. It can be thought that each impurity represents a charge, and with this, a potential is assigned and it is this potential that is responsible for the electrical scattering of the samples. Because of the impurities in the substrate, which can be negative or positive, regions with different concentrations and/or charge carriers are produced. In this way, it is common to find small islands of positive (-p-type) or negative (n-type) charges [1].

One way to get round the influence of the substrate on graphene properties is to produce suspended graphene samples. Suspension is a method in which the substrate used beneath the graphene is partially or completely removed. To produce these samples, the graphene is deposited on the SiO2 substrate and the metal contacts are arranged on the graphene flakes. The substrate is corroded on the surface below the graphene and so the graphene is suspended above the substrate and supported only by the metal contacts [3]. Using this method, the highest mobilities observed in graphene were obtained [15, 16, 17] and it was possible to make observations of the fractional quantum Hall effect in these suspended graphene samples [18, 19].

3 Materials and methods

3.1 Sample processes and procedures

This process, in which the layers are detached from each other (figure 3), is only possible because the interaction force between the layers is weak, i.e. of the Van Der Waals type. As a result, these layers are easily detached from the crystal when an external mechanical force is applied. *Scotch tape* can be used to apply this technique. The tape is then deposited on a silicon surface covered with a silicon dioxide dielectric. Using a soft object such as a finger, the tape is pressed down until the material adheres to the silicon oxide. Using an optical microscope [14], it is possible to identify some would-be graphene monolayers or bilayers due to the difference in contrast between the samples and the substrate. This difference in contrast is due to the difference in optical path between the region where there is or isn't graphene [1].

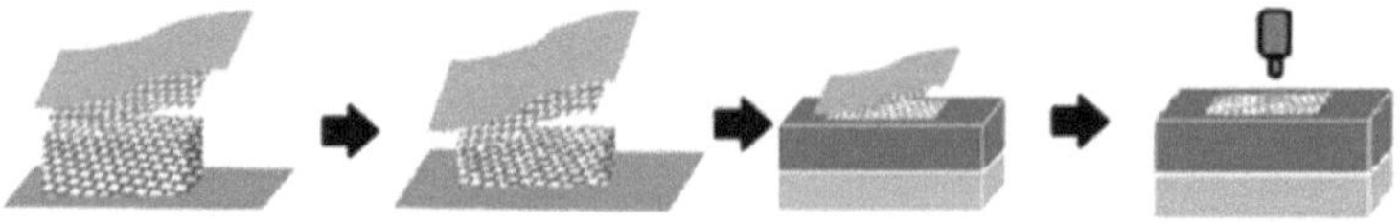

Figure 3Mechanical exfoliation process of the graphite flake, followed by deposition on the silicon dioxide substrate and observation of the sample under an optical microscope.

Through micro-exfoliation it is possible to obtain graphene flakes of up to 1mm. In addition, with this method we can achieve the best structural and electrical characteristics, as graphene has lower defect rates. However, despite preserving its structural quality (high mobility, absence of defects, etc.), this procedure is not suitable for large-scale production. Finally, for industrial purposes, the acquisition of larger flakes and deposition over large areas, other techniques are more effective [1].

3.2 Sample preparation:

To prepare the sample, the substrates (Si/SiO2 - 300 mm - ρ: 0.005~0.010Ω - no cleaning) were first cut into strips. A pen with a diamond tip was used and the cut was made in the cleavage plane). Once the marking was made on the substrate (figure 4,5), break it according to the cut.

Figure 4Profile of the cut made by the pen in the silicon dioxide substrate.

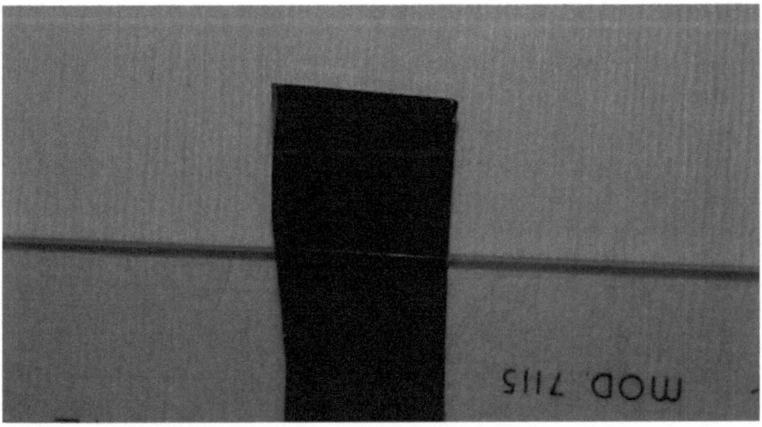

Figure 5Surface of the silicon dioxide with the mark made by the pen.

The sample was produced using a method called Mechanical Cleavage [22]. This method consists of separating the graphene layers that are stacked on a graphite crystal. This procedure is carried out with the aid of an adhesive tape and by sticking and unsticking the tape several times the interactions between the layers are broken (figure 6).

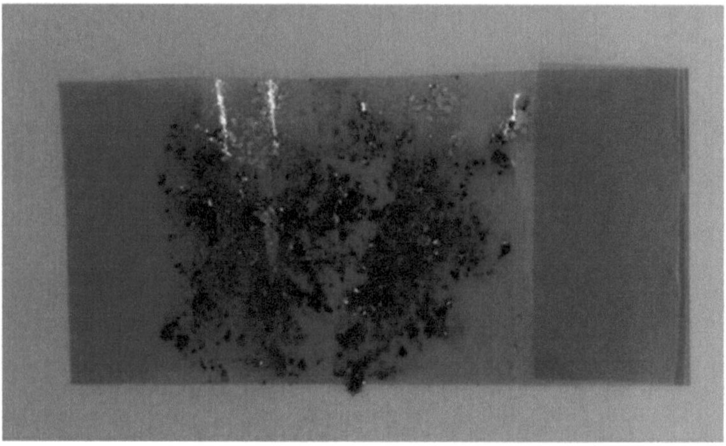

Figure 6Result of gluing and peeling along the adhesive tape.

Using a larger piece of tape, the previous adhesive tape was "stamped" over it (figure 7).

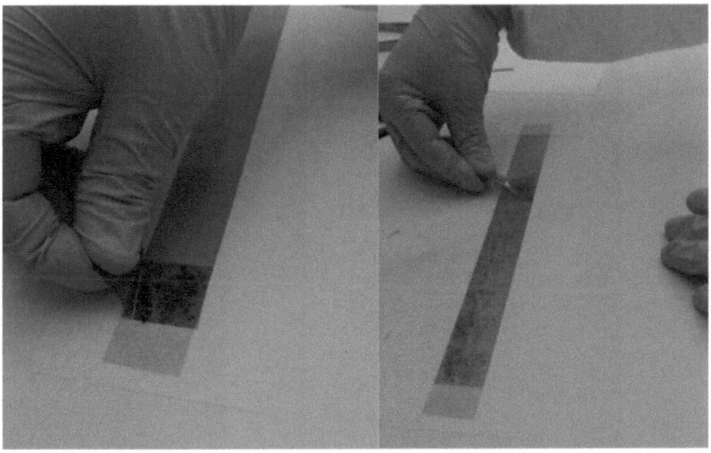

Figure 7Process of gluing and ungluing along a larger piece of adhesive tape in order to reduce the amount of graphite. The picture on the right shows the start of this process and the pictures on the left show the end. We can see from this picture on the left the reduced amount of graphite flakes.

This, in turn, will contain a smaller number of graphene layers and so it is attached to a silicon dioxide substrate (figure 8).

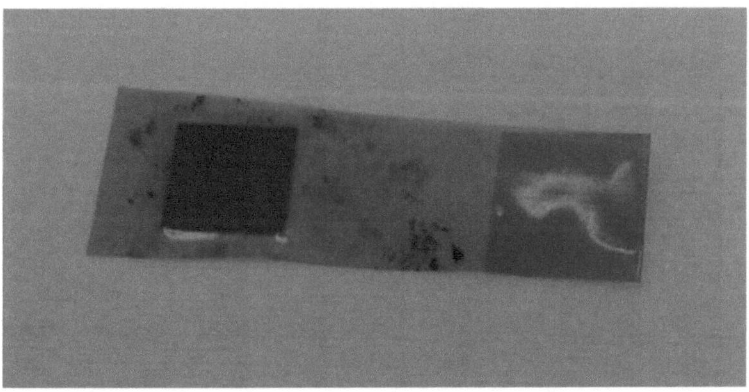

Figure 8Adhesive tape with a smaller amount of graphite flakes being deposited on the silicon dioxide substrate.

The contrast between the sample and the substrate is important for defining the number of layers of flakes (figure 9).

Figure 9Contrast observed to define the number of graphene layers.

The silicon dioxide substrate is uncleaned and acetone and isopropyl alcohol were used to clean it. The substrate was agitated in the ultrasonicator for 5 minutes (figure 10). The excess reagents were then removed using nitrogen gas (figure 11). The importance of this cleaning process is to improve the adhesion of the graphene to the silicon dioxide.

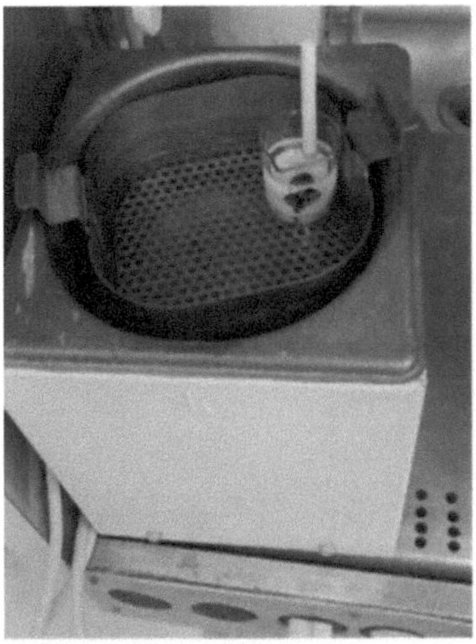

Figure 11Ultrasound agitating a beaker containing acetone and the SiO2/Si substrate.

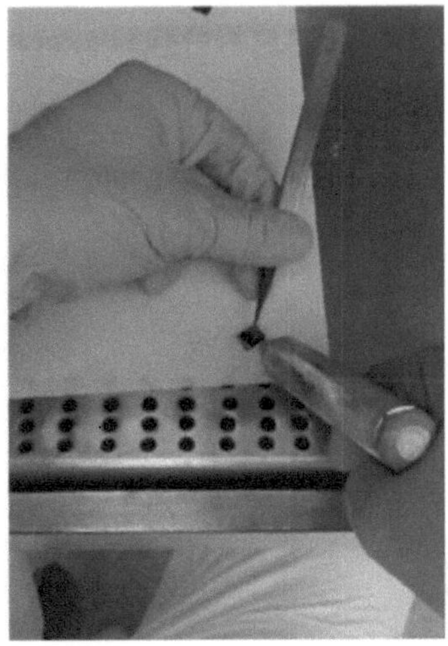

Figure 10Use of nitrogen gas to remove excess reagents.

4 Manufacturing and Characterisation

4.1 Device manufacture

A photolithography technique was used to fabricate the electrical contacts in the graphene and in the few graphene layers. A material sensitive to ultraviolet light was used to protect the surfaces of the samples. In this work, S1813 photoresist was used. A drop of the photoresist was deposited on the sample and then the sample was taken to a device on which it was rotated. This rotating base (figure 12, 13), which is called a *spinner*, causes the photosensitive material to form a thin, uniform film. The base was set to rotate with the following specifications: 1000 RPM - 5 s / 8000 RPM - 45 s.

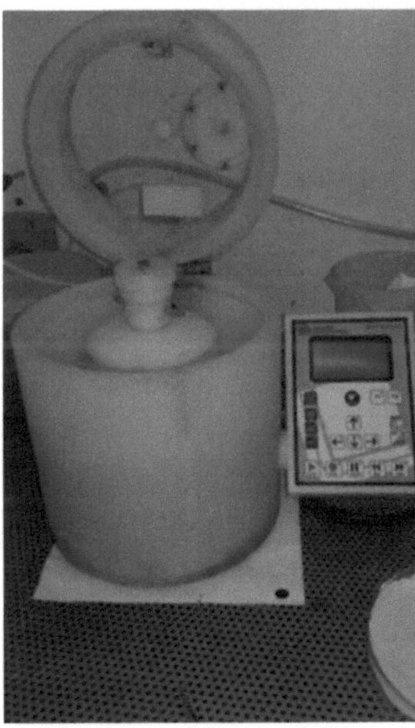

Figure 13Rotating base or Spinner.

Figure 12Sample positioned on the rotating base.

The sample was then taken for a curing process on the *hotplate* (figure

14), a hot plate, at a temperature of 100°C for 60 seconds.

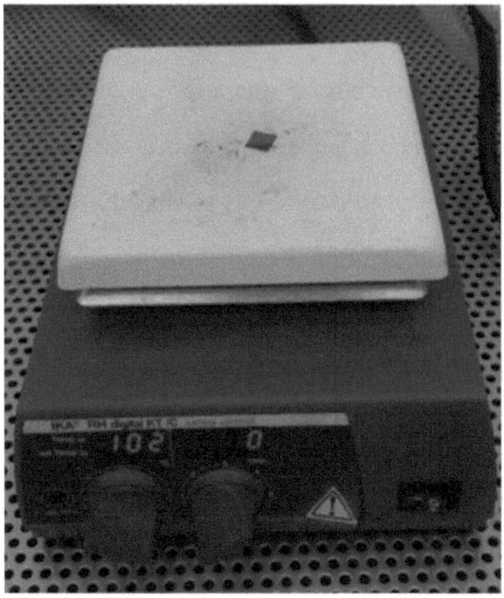

Figure 14Hotplate where the photoresist curing process takes place.

After curing the photoresist, the sample is transferred to the glass slide which will be positioned in the laser direct writing lithography equipment (Microtech LW405) and exposing the photoresist according to the pattern previously created in CAD software.

In the lithography equipment, a laser (wavelength 405 nm) is focussed through a lens and moves along the x-axis while the sample moves along the y-axis. Using the command to make the movements along the x and y axes and also controlling the intensity of the beam, the laser draws the submitted pattern onto the photoresist. At the end of this stage, the sample is placed in a solution containing an appropriate developer, and in this case we used MF26A developer for 35s.

According to the pattern implemented on the photoresist, Cr and Au were evaporated, respectively. The better adhesion of Au to the substrate is due to Cr [23]. To remove the excess photoresist, we adopted a process known as *lift-off* [24] (figure 15) in which we immersed the sample in acetone, leaving only the Cr/Au contact.

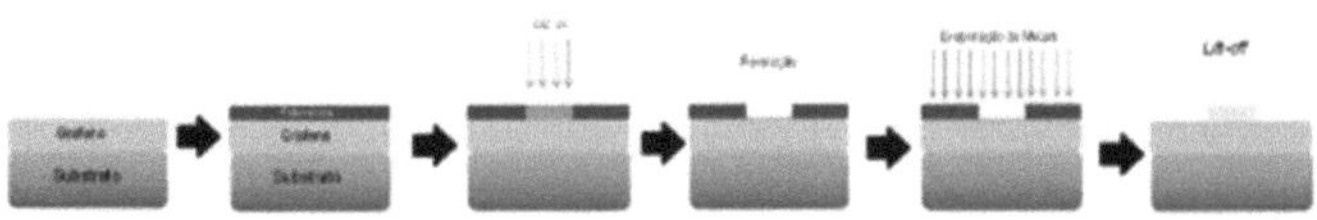

Figure 15Construction process for electrical contacts and *lift-off.*

4.2 Structural characterisation

Raman spectroscopy was used for this type of characterisation [25]. Figure 16 was produced in the Raman Spectroscopy Equipment at the Federal University of Viçosa, which allowed us to assess the quality of the sample. This technique consists of shining a laser beam on the sample, which in turn interacts with the sample and is then scattered (inelastic scattering). The information from this scattering is analysed [26]. Using the so-called *tight binding* model [27,28] we can calculate the electronic dispersion of graphene, as this provides the information needed to understand its electronic structure and the irregularities in its crystal lattice. By mapping it across the sample, we can analyse its quality and observe its irregularities. Figure 17 shows the peaks known as the G band, the G' band and the D band.

The D band is related to the disorder or defects in the crystalline network of the material, graphene [29]. The G band comes from the tangential vibrational modes of the graphene surface and from it we can extract information about whether the sample is doped or undoped. The G' band, or 2D band, is the result of the second-order resonance of the D band, which is linked to its excitation energy [29]. The G' band determines whether the graphene analysed is a monolayer or a multilayer.

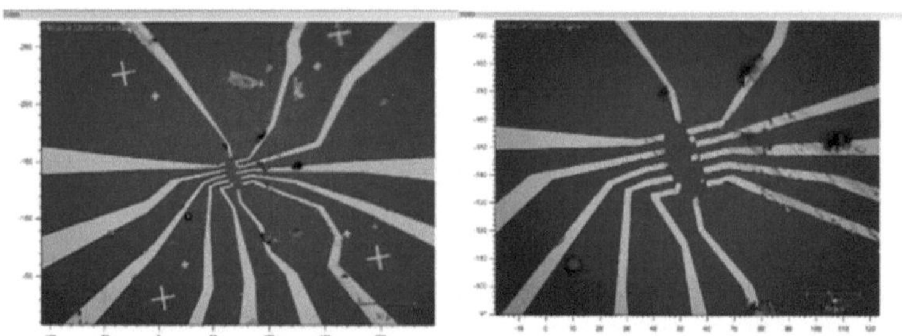

Figure 16 Left side: 20x Raman spectroscopy image. Right side: 50x Raman spectroscopy. Images provided by the Raman Spectroscopy Laboratory at the Federal University of Viçosa.

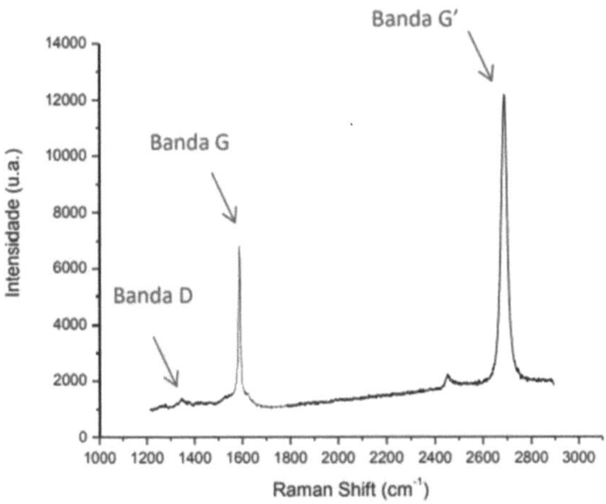

Figure 17Raman spectrum of the graphene sample with Cr/Au deposition.

4.3 Electrical characterisation at low temperatures

The material's conduction properties are measured at low temperatures using a cryostat (figure 18). This equipment allows measurements to be made at low temperatures (10 K). Its cooling comes from the high pressure that the gas is subjected to and as the gas moves along the cryostat stages it cools down, removing heat from the cold finger (figure 18). In a compressor, the gas is trapped and through an internal procedure the gas is compressed and transferred to the cold finger via a flexible hose. To reuse it, a flexible hose is used to return it to the compressor. This cycle continues for as long as it takes to reach the desired temperature. The sample is placed inside the cold finger, which will be covered by a cylinder coupled to a vacuum pump that will be subjected to a high vacuum (10-6 torr). A resistor wrapped around the cold finger provides heating in the structure where the sample is located and two silicon diode sensors are used to control the temperature. One of them is inserted into the cold finger and the other where the sample has been deposited. To control the temperature, the *LakeShore 331 Temperature Controller* (figure 18) is used, which is connected to a computer via the GPIB (*General Purpose Interface Bus*) output, where it will receive the information as programmed. The *Keithley 2400* (figure 18), which is capable of operating as both a current and voltage source, is coupled to the system and its purpose is to take electrical measurements.

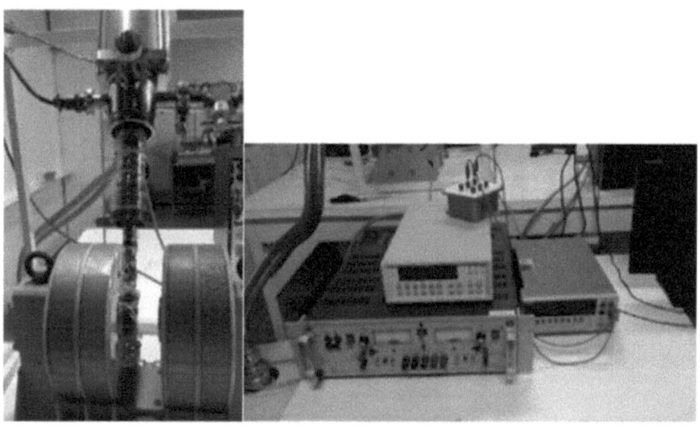

Figure 18Left side: Cryostat component, called the cold finger. Right side: ***LakeShore 331 Temperature Controller*** and ***Keithley 2400.***

4.4 Morphological characterisation:

To carry out this type of characterisation, we used the technique of atomic force *microscopy* (AFM), which consists of measuring forces or interactions between a tip and the surface of the sample. Through these interactions, the morphology of the surface under analysis can be mapped.

The basic schematic of the AFM is shown in figure 19. The *cantilever* is a rod on which one end is fixed to a base and the other end is fitted with a tip (or probe). As the sample is scanned, the *cantilever* deflects due to the interaction of the force between the atoms in the sample and the atoms in the tip. If the force is attractive, the *cantilever* tilts towards the sample and if the force is repulsive, it tilts away from the sample. The distance between the sample and the tip is controlled by the *software* or manually. Once the distance has been set, the laser is focused on the rod and as it undergoes deflections due to the interaction between the sample and the tip, the laser is deflected and captured by a photodiode. The photodiode transmits the information about the deviation suffered by the rod to a detector, which in turn transmits it to a computer that builds topographical images of the sample's surface.

The image can be formed in three ways: contact, semi-contact and non-contact. The image formed in this work is due to the semi-contact or *tapping* mode. This mode consists of the rod oscillating almost at its resonance frequency, with a high amplitude and its contact remaining periodically alternating with the surface of the sample [32]. The oscillation amplitude can be controlled by the average distance between the sample and the tip. This mode is used to avoid damaging the sample.

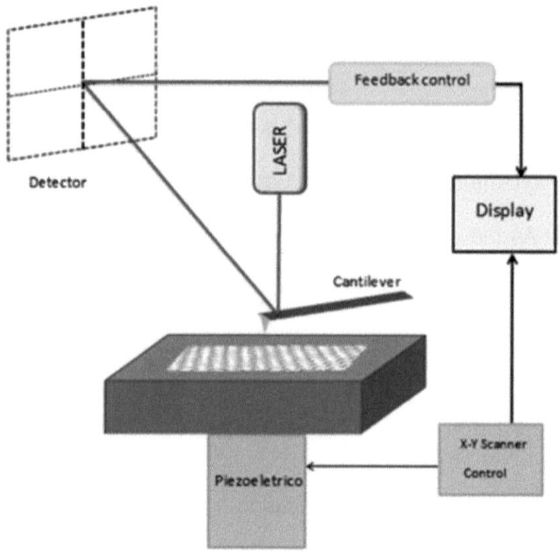

Figure 19Basic schematic of the AFM.

5 Results and Discussion

This work involves the manufacture of electronic devices based on mono-, bi- and trilayer graphene, following the procedures described below.

1st) Exfoliation of the material on a substrate, such as SiO_2;

The exfoliation of the material on the silicon dioxide substrate was done manually under the supervision of Dr Eduardo Nery at the Federal University of Minas Gerais. The substrate was first cleaned, as explained in the text PREPARATION OF SAMPLES. This process is important because impurities such as dust, gases adsorbed on the surface of the substrate and moisture can influence and/or hinder the observation, adhesion or quality of the sample.

2nd) Detection of mono-, bi- or trilayers. In the case of graphene, optical microscopy is sufficient, but other materials require Raman spectroscopy or atomic force microscopy (AFM) to be localised at their ultrafine scale;

Detection is a very detailed process and sometimes only with the help of optical microscopy can you detect a mono, bi or trilayer of graphene. This observation process can be time-consuming, because no matter how clean the substrate is, some impurities can adhere to it, or because of the type of adhesive tape used, glues can remain on the substrate that can be mistaken for the material. At UFMG, the graphene was only observed using an optical microscope and the other analyses, such as Raman spectroscopy and AFM microscopy, were carried out at UFV. It is worth mentioning that the other procedures for manufacturing the device were carried out at UFMG and once it was ready, it was analysed at the UFV Physics Department.

The sample was first analysed using the AFM microscope at the UFV Physics Department (figure 20). The basic operation of the AFM was described in the text MORPHOLOGICAL CHARACTERISATION. In this procedure we used the semi-contact mode so as not to damage the sample and the result we obtained was a topographic image of the device. Using the programme provided, we can extract the dimensions of the graphene and the electrical contacts that were produced and, using this data, we can get an idea of the size of the device manufactured. Figure 20 shows the topography of the device and Figures 21, 22 and 23 show the dimensions of the graphene and the size of the electrical contact.

Figure 20Image generated by the Nova programme in the AFM Laboratory of the UFV Physics Department.

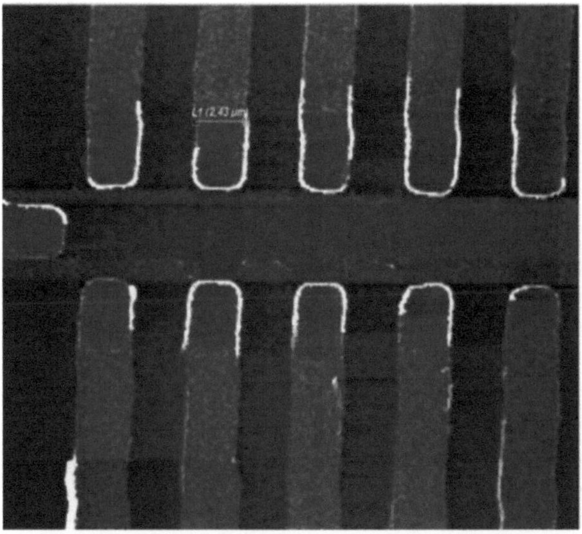

Figure 21Image generated by the Nova programme in the AFM Laboratory of the UFV Physics Department showing the width of the electrical contacts.

Figure 22Image generated by the Nova programme in the AFM Laboratory of the UFV Physics Department showing the length of the graphene.

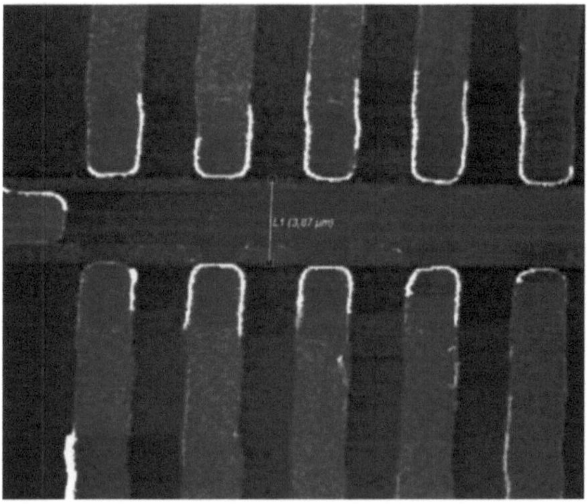

Figure 23Image generated by the Nova programme in the AFM Laboratory of the UFV Physics Department showing the width of the graphene.

The analysis carried out using Raman spectroscopy tells us about the properties of graphene, such as: the disorder or defects of its crystalline network throughout its length, whether the sample has been doped or not and whether the graphene is a monolayer, a biolayer or a trilayer. These results are shown in graph 1, which was obtained by mapping the entire length of the graphene. By shifting the D band ~1348 cm^{-1} , we can see that there is a disorder along its crystalline network, as there is a contrast between the peaks in this region. There is a single peak in the G band at ~1585 cm^{-1} and comparing this with the value obtained in the graph shows that the sample analysed is not doped. Finally, the G' band peak tells us the number of graphene layers used. This peak, which is at ~2686 cm^{-1} , compared with the values previously given for graphene, means that the device manufactured is a graphene monolayer.

These analyses help us to obtain information about the material under study, because when the procedures for manufacturing the device were carried out at UFMG, we only had a brief idea of the results. Through these analyses we obtained general information about the material.

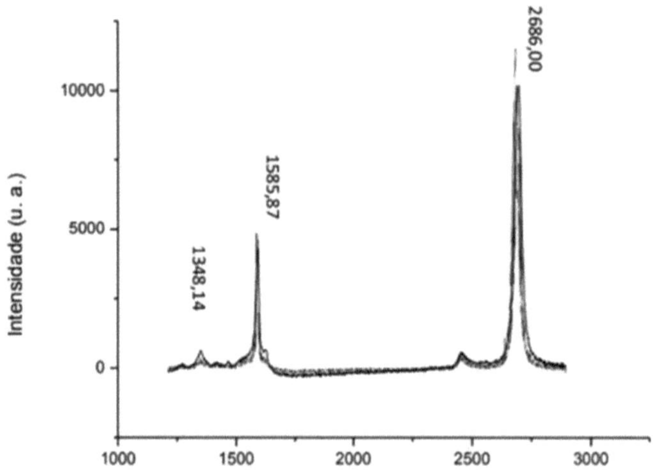

Graph 1Raman spectrum with Cr/Au deposition on the electrical contacts.

3rd) Lithography using a light beam, via *laserwriter*;

Lithography was carried out at UFMG using LW405 equipment. A 405 nm wavelength laser is responsible for the pattern drawn on the device. This equipment was run by Dr Eduardo Nery, as prior training is required to operate it. As described in the text MANUFACTURING THE DEVICE, the pattern formed must be drawn on the photoresist so as not to damage the sample. Care must be taken so that the sample is not lost after a long labour to make it.

4th) Deposition of chrome and gold for *source-drain* contacts;

The deposition of Cr and Au on the pattern produced by the lithography equipment will serve to make the electrical contact and carry out the electrical measurements. We evaporated 3 nm of Cr and then 30 nm of gold. We used Cr before Au so that it would adhere better to the pattern drawn on the substrate. The contacts produced play the role of the "positive and negative" poles of the circuit, i.e. where the current "enters and leaves" the sample.

5th) *"Lift-off"* of the samples, to remove resistors and spurious metals/oxides.

The *lift-off* is important because it finalises our sample. We can compare it to something that is dirty and then becomes clean after a bath. To do this, the sample is dipped in acetone and the result is that the photoresist is removed and only the Cr/Au pattern remains (figure 24). This finalises the sample manufacturing procedure.

Figure 24AFM image of the Cr/Au electrical contact.

6 Conclusions and prospects

Finally, the experience gained in the process of fabricating devices in two-dimensional materials is emphasised, including electron beam lithography, metal deposition, morphological and spectral characterisation. The sample fabrication process was carried out at the Physics Department of UFMG and the morphological and spectral analyses were carried out respectively at the AFM and Raman Labs of the Physics Department of UFV.

The aim is also to implement the manufacture of electrical devices, including the exfoliation process, mono, bi or trilayer detection, lithography, metal deposition and *lift-off* at the Physics Department of the Federal University of Viçosa.

7 Bibliographical references:

[1] - CADORE, Alisson R.; *"Influence of Ionised Impurities and Adsorbed Molecules on the Electrical Transport Mechanism of Graphene"*; Universidade Federal de Minas Gerais; Belo Horizonte, Minas Gerais, Brazil, 2013;

[2] - GUIMARÃES, Marcos H. D.; *"Electronic and structural aspects of graphene and derivatives: a theoretical-experimental study"*; Universidade Federal de Minas Gerais; Belo Horizonte, Minas Gerais, Brazil, 2010;

[3] - BRANT, Juliana C.; *"Electrical transport in graphene nanostructure: influence of functionalisation, geometry and substrate doping"*; Universidade Federal de Minas Gerais; Belo Horizonte, Minas Gerais, Brazil, 2011;

[4] - Saito, R., Dresselhaus, G., and Dresselhaus, M. S.; *"Physical Properties of Carbon Nanotubes"*; Imperial College Press, 1st edition; (1998).

[5] - Malard, L. M., Guimarães, M. H. D., Mafra, D. L., Mazzoni, M. S. C., and Jorio, A.; Physical Review B, 79 (12); 125426; 2009;

[6] - Geim, A. K. and Novoselov, K. S.; Nature Materials; 6(3), 183-191; 2007;

[7] - Stankovich, S., Dikin, D. A., Piner, R. D., Kohlhaas, K. A., Kleinhammes, A., Jia, Y., Wu, Y., Nguyen, S. T., and Ruoff, R. S.; Carbon; 45(7), 1558 - 1565; 2007;

[8] - Dikin, D. A., Stankovich, S., Zimney, E. J., Piner, R. D., Dommett, G. H. B., Evmenenko, G., Nguyen, S. T., and Ruoff, R. S.; Nature; 448(7152), 457-460; 2007;

[9] - Mkhoyan, K. A., Contryman, A. W., Silcox, J., Stewart, D. A., Eda, G., Mattevi, C., Miller, S., and Chhowalla, M.; Nano Letters; 9(3), 1058-1063; 2009;

[10] - Sofo, J. O., Chaudhari, A. S., and Barber, G. D.; Physical Review B; 75(15); 153401; 2007;

[11] - Elias, D. C., Nair, R. R., Mohiuddin, T. M. G., Morozov, S. V., Blake, P., Halsall, M. P., Ferrari, A. C., Boukhvalov, D. W., Katsnelson, M. I., Geim, A. K.,and Novoselov, K. S.; Science; 323(5914), 610-613; 2009;

[12] - Du, X., Skachko I. and Andrei E. Y.; *Inter. J. of Modern Physics B*, 22 (25/26), 4579-4588; 2008;

[13] - Novoselov, K.S. *et al*; Science; 306, 666- 669; 2004;

[14] - Blake, P. et. al.; *App. Phys;* Lett. 91, 063124; 2007;

[15] - Bolotin, K., Sikes, K., Jiang, Z., Klima, M., Fudenberg, G., Hone, J., Kim, P. and Stormer, H.; *Solid State* Comm, 146; 2008;

[16] - Bolotin, K., Sikes, K., Hone, J., Stormer, H. and Kim, P.; *Phys. Rev. Lett.*, 101; 2008;

[17] - Du X., Skachko, I., Barker, A. and Andrei, E.Y.; *Nature Nano*, 3; 2008;

[18] - Du, X., Skachko, I., Duerr, F., Luican, A. and Andrei, E.Y.; *Nature* 462; 2009;

[19] - Bolotin, K.I., Ghahari, F., Shulman, M.D., Stormer, H.L. and Kim, P.; *Nature*, 462, 2009;

[20] - Ghahari, F., Zhao, Y. P., Cadden-Zimansky, Bolotin, K. and Kim, P.; *Phys. Rev. Lett.* 106; 2011;

[21] - "*Class for Physics of the Royal Swedish Academy of Sciences*"; http://www.nobelprize.org/nobel_prizes/physics/laureates/2010/advanced-physicsprize2010.pdf; 2015;

[22] - S. Novoselov, K., K. Geim A., V. Morozov S., Jiang D., Zhang Y., V. Dubonos S., V. Grigorieva I. and A. Firsov A., *Science*, **306**, 666 (2004).

[23] - R. Miquita D., *Fabrication and Testing of Semiconductor Devices*, graduation monograph, Universidade Federal de Viçosa, Viçosa, MG, Brazil (2001).

[24] - E Willians R., *Gallium Arsenide Processing Techniques*, Artech House, Inc. Dedham (1985).

[25] - C. Ferrari A., C. Meyer J., Scardaci V., Casiraghi C., Lazzeri M., Mauri F., Piscanec S., Jiang D., S. Novoselov K., Roth S. and K. Geim A., *Phys. Rev. Lett.*, **97,** 187401 (2006).

[26] - V. Raman C. and S. Krishnan K., *Nature*, **121**, 501 (1928).

[27] - Saito R., Dresselhaus G. and S. Dresselhaus M., *Physical Properties of Carbon Nanotubes*, Imperial College Press (London), 1998.

[28] - Reich S., Thornsen C. and Maultzsch J., *Carbon Nanotubes Basic Concepts and Physical Properties*, Berlin, Germany, 2004.

[29] - DOS SANTOS, Fabrício A.; *"Processing of Oxidised Graphene in the form of ultra-thin films and application in sensors"*; São Carlos School of Engineering; São Carlos Institute of Chemistry of the University of São Paulo; São Carlos, São Paulo, Brazil, 2012.

[30] - Wallace, P. R.; "*The Band Theory of Graphite";* Phys. Rev. 71, 622 .

[31] - K. Lu X., F. Yu M., Huang H., and R. Ruoff S.; *Nanotechnology* ;10; 269; 1999.

[32] - Pinto P. E., Ramos, Q. G. Fonseca Filho D. H.; *The Atomic Force Microscope (AFM): an important tool in the study of surface morphology on the manometric scale;*

<https://periodicos.unifap.br/index.php/estacao/article/viewFile/1167/ervetonv3n2.pdf>

Figures:

[23] http://cnx.org/contents/f06226c5-c2a4-4798-9c75 b016acea73cd@2/Characterization-of-Graphene-b; 2015;

[24] - http://spectrum.ieee.org/semiconductors/materials/graphene-makes-transistors-tunable; 2015.

MIX
Papier aus verantwortungsvollen Quellen
Paper from responsible sources
FSC® C105338

Printed by Books on Demand GmbH, Norderstedt / Germany